SWCS PUB 09-1, NOV 2009

A Leader's Handbook to Unconventional Warfare

LTC Mark Grdovic

U.S. ARMY JOHN F. KENNEDY
SPECIAL WARFARE CENTER AND SCHOOL

Table of Contents

U.S. Army John F. Kennedy Special Warfare Center and School	2
Foreword	6
About the Author	7
Understanding Unconventional Warfare	9
The Conduct of UW	10
The Guerrillas	10
The Underground	11
The Auxiliary	12
Resistance Force Commands	13
How the Environment Shapes the Structure of an Insurgency	13
Phases of an Insurgency	14
Development of a Resistance Movement Following an Invasion	16
How the U.S. Military Conducts UW	17
UW in Support of General War	17
UW in Support of Limited War	20
The Roles of the Army, Air Force, Navy and Interagency	22
Prerequisite Conditions for Successful UW	22
A Weakened or Unconsolidated Regime	23
A Population of Strong Character	23
Favorable Terrain	23
Capable and Skilled Indigenous Leadership	24
The Criticality of the Feasibility Assessment	25
Risk Analysis and Risk Acceptance	26
Logistics Challenges	27
Command and Control Challenges	27
Preparatory Measures Taken Within an Allied Nation at Risk	29
Preparatory Measures Focused Toward a Potential Adversary	30
The Impact of Technology on UW	30
Law of Land Warfare Challenges	31
U.S. UW Efforts from 1951-2003	32
U.S. Military and CIA in Korea (1951-1953)	32
CIA in Albania and Latvia (1951-1955)	32
CIA in Guatemala (1954)	33
CIA in Tibet (1955-1969	33
CIA in Indonesia (1957-1958)	34
CIA in Cuba and the Bay of Pigs (1961)	34
CIA and Special Forces in Laos (1959-1962)	34
CIA and Special Forces in North Vietnam (1961-1964)	35
South Vietnam (1957-1975)	35
CIA and Special Forces in Nicaragua and Honduras (1980-1988)	36

Table of Contents (Contd.)

CIA and Special Forces in Pakistan and Afghanistan (1980-1991) ... 36
Cold War Contingency Plans for Scandinavia and Europe (1952-1989) ... 37
Kuwait (1990-1991) ... 37
Afghanistan (2001 - 2002) ... 37
Iraq (2002-2003) ... 38

U.S. Army John F. Kennedy Special Warfare Center and School

The U.S. Army John F. Kennedy Special Warfare Center and School, or SWCS, is responsible for the training, personnel, doctrine and policy to support Army Special Operations Forces, or ARSOF. SWCS serves as the U.S. Army Special Operations Command, or USASOC, proponent for all matters pertaining to individual training, develops doctrine and all related individual and collective training material, provides leader development, develops and maintains the proponent training programs and systems, and provides entry-level and advanced individual training and education for Civil Affairs, Psychological Operations and Special Forces. SWCS, a component subordinate command of USASOC, constitutes the training center and institution of ARSOF.

The Directorate of Special Operations Proponency has the responsibility for managing the careers of all Civil Affairs, Psychological Operations and Special Forces Soldiers from the time they enter one of the career fields until they leave.

The SWCS' Directorate of Training and Doctrine develops all Army special-operations doctrine and works with the field and trainers at SWCS to develop all courses and training programs. The SWCS has some of the most up-to-date doctrine in the Army, with 30 field manuals and other doctrinal publications being revised or written at any given time.

The Directorate of Special Operations Education serves as the USAJFKSWCS adviser for education issues. It provides training and education opportunities that enable the USAJFKSWCS staff and faculty to perform as flexible, adaptive ARSOF leaders and trainers. Key in the training of special-operations forces is language and culture training, which falls under the auspices of the Directorate of Special Operations Education.

The Joint Special Operations Medical Training Center is responsible for all U.S. military special-operations forces combat medical training including Army Rangers, Civil Affairs, Special Forces and Navy SEALs.

Advanced education for Army special-operations Soldiers is provided through the Noncommissioned Officer Academy and the Warrant Officer Institute. The NCO Academy prepares enlisted Soldiers for leadership positions in all ARSOF including Civil Affairs, Psychological Operations and Special Forces. Soldiers receive training in leadership skills, military studies, resource management, effective communication, operations and intelligence, unconventional warfare, operational planning, psychological operations, civil-military operations and advanced military occupational skills.

The Special Forces Warrant Officer Institute educates and trains warrant officer candidates, newly appointed warrant officers, and senior warrant officers at key points in their career in order to provide the force with competent combat leaders and planners.

The Center and School's Training Group, consisting of five battalions, conducts the complete spectrum of special-operations training from the Civil Affairs and Psychological Operations advanced individual training to CA, PO, and the SF Qualification Course.

The Army Special Operations Capabilities Integration Center anticipates the future environment, threats, and requirements for the ARSOF, and to look at what can be done today to prepare for tomorrow's issues. The center validates concepts through experimentation, war-gaming and formulates the impact on proposed ARSOF futures capabilities and operational architectures.

The International Military Student Office is responsible for the oversight of international students who attend any one of many courses offered by the SWCS.

The Center and School conducts more than 44 different courses and trains more than 14,000 students annually. Since 1963, SWCS has led the way in creating the world's finest special operations forces.

A Leader's Handbook to Unconventional Warfare

LTC Mark Grdovic

U.S. ARMY JOHN F. KENNEDY
SPECIAL WARFARE CENTER AND SCHOOL

Comments about this publication are invited and should be forwarded to Director, Directorate of Training and Doctrine, U.S. Army John F. Kennedy Special Warfare Center and School, 3206 Ardennes St., Fort Bragg, N.C. 28310. Copies of this publication may be obtained by calling SWCS at 910-432-5703. Electronic versions of this publication can be found at www.arsofu.army.mil.

The Special Warfare Professional Development Publication Office is currently accepting works relevant to Army special operations for potential publication. For more information, please contact Mr. Jerry Steelman or Mrs. Janice Burton, Special Warfare at 910-432-5703 or Steelman@soc.mil or Burtonj@soc.mil.

This work has been cleared for public release, distribution is unlimited.

The views expressed in this publication are entirely those of the author and do not necessarily reflect the views, policies or positions of the United States Government, the Department of Defense, the United States Special Operations Command, the United States Army Special Operations Command or the U.S. Army John F. Kennedy Special Warfare Center and School.

Foreword

At the end of World War II, the Army developed the concept of unconventional warfare, or UW, based largely on the experiences of Soldiers who had worked with resistance movements during the war. The concept was formally introduced into doctrine in 1955, specifically to convey a wider responsibility than simply working alongside guerrilla forces. UW is currently defined as activities conducted to enable a resistance movement or insurgency to coerce, disrupt or overthrow a government or occupying power by operating through or with an underground, auxiliary or guerrilla force in a denied area. From the beginning, UW has been a mission of Special Forces, and the JFK Special Warfare Center and School has been the proponent for UW training and doctrine.

Our operating environment has been less constant since the 1950s, however. We no longer face the threat of major combat operations, as during World War II. The dominant form of warfare that now confronts the United States and will likely do so for the remainder of the 21st century is irregular warfare. The five components of irregular warfare — counterterrorism, foreign internal defense, stability operations, counterinsurgency and UW — will increasingly involve all elements of the U.S. military and other elements of national power. UW thus presents a challenge to conventional forces and special-operations forces alike.

This paper, although it is not doctrine, is intended to introduce a new generation of Soldiers to the concept, the history and the techniques of UW. It is intended to be the first in a series of papers published to inform readers and provoke thought on a number of topics that are of interest to members of the special-operations community and, in some cases, general-purpose forces as well.

Major General Thomas Csrnko
Commander
U.S. Army John F. Kennedy Special Warfare Center and School

About the Author

Lieutenant Colonel Mark Grdovic is assigned to the U.S. Special Operations Command Central. His previous assignment was as the director of the USASOC G3X Special Program Division. Other assignments include Director of the President's Emergency Operations Center, White House Military Office. He was formerly chief of the Special Forces Doctrine Branch, SF Doctrine Division, in the U.S. Army John F. Kennedy Special Warfare Center and School's Directorate of Training and Doctrine.

His other SF assignments include service with the 1st Battalion, 10th SF Group, as commander of SF detachments 016 and 032; small-group instructor for the officer portion of the Special Forces Qualification Course; company commander and S3, 3rd Battalion, 10th SF Group; and commander, Company A, 4th Battalion, 1st Special Warfare Training Group. Lieutenant Colonel Grdovic holds a bachelor's degree from SUNY Cortland University and a master's degree from King's College London.

In 2002, Grdovic was one of the main architects of the UW campaign plan for Northern Iraq as part of Operation Iraqi Freedom. In 2003, he infiltrated northern Iraq prior to the start of the invasion in order to link-up with the leadership of the Patriotic Union of Kurdistan (PUK) and synchronize the Kurdish efforts. In 2007-2008, he served as the CJSOTF-AP LNO to Multi- National Force Iraq (MNF-I).

Lieutenant Colonel Mark Grdovic, far right, was one of the main architects of the UW campaign plan for Northern Iraq as part of Operation Iraqi Freedom.

Understanding Unconventional Warfare

Unconventional warfare, or UW, is defined as activities conducted to enable a resistance movement or insurgency to coerce, disrupt or overthrow an occupying power or government by operating through or with an underground, auxiliary and guerrilla force in a denied area. Inherent in this type of operation are the inter-related lines of operation of armed conflict and subversion. The concept is perhaps best understood when thought of as a means to significantly degrade an adversary's capabilities, by promoting insurrection or resistance within his area of control, thereby making him more vulnerable to a conventional military attack or more susceptible to political coercion. "UW includes military and paramilitary aspects of resistance movements. UW military activity represents the culmination of a successful effort to organize and mobilize the civil populace against a hostile government or occupying power. From the U.S. perspective, the intent is to develop and sustain these supported resistance organizations and to synchronize their activities to further U.S. national security objectives." [1]

In many regards UW is a unique and appropriately unpublicized category of [instability] operations that complements the Army's full-spectrum operations consisting of offense, defense and stability operations. The successful conduct of UW can significantly contribute to a conventional military campaign or even provide an alternative option when the application of conventional military power is not appropriate. However, conducting UW is not applicable to all situations, environments or adversaries. The promotion of instability into an environment is not always beneficial to the long-term strategy for a region. There are certain prerequisite conditions that must be present in order for a UW effort to be successful. These conditions (sometimes referred to as the UW potential) cannot be artificially manufactured if they do not exist within the environment. When viewed in this context, the potential strategic and operational value, as well as the potential risks, becomes more apparent.

While many of the tactics and techniques utilized within the conduct of UW have significant application and value in other types of special operations, many of these techniques, such as sabotage or intelligence collection, are not exclusive to UW, and subsequently should not be categorized as such when conducted as part of other special operations. Similarly, the technique of utilizing indigenous forces or surrogates is a methodology applicable to any type of special operation. In 1943, the Allied Forces enlisted the assistance of the Norwegian Resistance to aid the Special Operations Executive (SOE) in their sabotage mission to destroy the Nazi's heavy-water facility in Norway. In 1998, selected elements from Afghanistan's Northern Alliance were employed to assist U.S. special operatives in capturing Osama bin Laden. While both of these examples represent the employment of indigenous irregular forces in special operations, they are more accurately direct action and counterterrorism rather than UW. Operations are more clearly categorized by what they intend to achieve rather than by individual techniques or who is conducting them.

Like all strategic capabilities, the value of UW is not measured by its frequency of employment, but by its potential effects. It should also be noted that UW represents only one of several types of special operations listed as U.S. Special Operations Command's responsibilities within Title 10 of the U.S. Code.

A Leader's Handbook to Unconventional Warfare

The Conduct of UW

Enabling a resistance movement or insurgency entails the development of guerrilla forces and an underground, both with their own supporting auxiliaries. The end result comes from the combined effects of "armed conflict," conducted predominantly by the guerrillas, and "subversion," conducted predominantly by the underground. The armed conflict, normally in the form of guerrilla warfare, reduces the host-nation's security apparatus and subsequent control over the population. The subversion undermines the government's or occupier's power by portraying them as illegitimate and incapable of effective governance in the eyes of the population.

In order for an adversary to be susceptible to the effects of insurgency or resistance, he must have some overt infrastructure and legitimacy that are vulnerable to attacks (physical or psychological). In this respect the "target recipient" does not have to be a state government, but it does have to possess state-like characteristics (e.g. the elements of national power), such as those similar to an occupying military force exercising authority through martial law. It is not uncommon for military planners to overfocus on the quantifiable aspects related to the more familiar armed conflict and underfocus on the somewhat less quantifiable and less familiar aspects of subversion.

While different insurgencies are unique to their own strategy and environment, U.S. Army Special Forces has long utilized a doctrinal construct that serves as an extremely useful frame of reference for describing the actual configuration of a resistance movement or insurgency. This construct depicts three components common to insurgencies: the guerrillas, the underground and the auxiliary. Without an understanding and appreciation for the various components (guerrillas, underground and auxiliary) and their relation to each other, it would be difficult to comprehend the whole of the organization based solely on what is seen. In this regard, insurgencies are sometimes compared to icebergs, with a distinct portion above the surface (the guerrillas) and a much larger portion concealed below the surface (the underground and auxiliary).

The Guerrillas

Guerrillas are the overt military component of a resistance movement or insurgency. As the element that will engage the enemy in combat operations, the guerrilla is at a significant disadvantage in terms of training, equipment and fire power. For all his disadvantages, he has one advantage that can offset this unfavorable balance...*the initiative*. In all his endeavors, the guerrilla commander must strive to maintain and protect this advantage. The guerrilla only attacks the enemy when he can generate a relative (albeit temporary) state of superiority. The guerrilla commander must avoid decisive engagements, thereby denying the enemy the opportunity to recover and regain their actual superiority and bring it to bear against his guerrilla force. The guerrilla force is only able to generate and maintain this advantage in areas where they have significant familiarity with the terrain and a connection with the local population that allows them to harness clandestine support.

A guerrilla force needs a base in order to rest, train, prepare and refit. The degree of sophistication of the guerrilla bases is proportional to the guerrillas' ability to establish an early-warning system (which includes support from the local population), the capability of the enemy counterguerrilla forces and the difficulty of the terrain. Bases need to be mobile enough to move if the supporting early-warning networks indicate the approach of a superior enemy force. The guerrilla also needs the ability to acquire, store and distribute large quantities of supplies without the benefit of standard lines of supply and communication. This is achieved through a decentralized network of caches across a wide area instead of maintaining large centralized stockpiles. This minimizes the loss of material if a base has to displace or is destroyed. It also allows the guerrillas to conduct operations across a wide area without a long logistics trail.

The guerrilla leader needs to communicate with other guerrilla bands, his auxiliary and possibly the "shadow government" residing inside the country or the resistance movement leadership or displaced government "in exile" residing outside the country, all within an area where the enemy forces are always actively looking and listening for any indicators that would compromise the location of his forces or supporting mechanism. The smaller the force, the less power the guerrillas can project. The larger the force, the more challenging it is to sustain the force without it becoming compromised. One of the greatest dangers to a guerrilla force is growing too quickly or beyond its means.

Depending on the degree of control over the local environment, the size of guerrilla elements can range anywhere from squad- to brigade-size groups. In the early stages of an insurgency, the guerrilla force's offensive capability might be limited only to small stand-off attacks. As the guerrilla force's base of support from the population grows, so does its ability to openly challenge government security forces with large-scale attacks. At some point in an insurgency or resistance movement, the guerrillas may achieve a degree of parity with the host-nation forces in certain areas. In these cases, units may start fighting more openly, sometimes referred to as partisans, rather than as guerrilla bands. In well-developed insurgencies, formerly isolated pockets of resistance activity may eventually connect and create liberated territories, possibly even linking with a friendly or sympathetic border state.

It is important to use the term "guerrilla" accurately in order to distinguish between other types of irregular forces that might appear similar, but are in fact something entirely different, such as militias, mercenaries or criminal gangs. The Department of Defense definition defines a guerrilla as someone who engages in guerrilla warfare. This is somewhat oversimplified, in that guerrilla warfare is generally considered a tactic that can be utilized by any force (regular or irregular). True guerrilla forces normally only exist, for an extended period of time, as part of a broader resistance movement or insurgency.

Counterinsurgency forces may utilize indigenous militias or intelligence networks that in many ways resemble parts of a resistance force. While these forces are "guerrilla-like," they operate in a much less restrictive environment than actual guerrillas and undergrounds. They are still part of the host-nation forces, and are employed in support of the goals of the governing authority or state.

The Underground

The underground is a cellular organization within the resistance movement or insurgency that has the ability to conduct operations in areas that are inaccessible to guerrillas, such as urban areas under the control of the local security forces. The underground can function in these areas because it operates in a clandestine manner, resulting in its members not being afforded legal belligerent status under any international conventions.

Examples of these functions include:
- Intelligence and counterintelligence networks
- Subversive radio stations
- Propaganda networks that control newspaper/leaflet print shops and or Web pages
- Special material fabrication (false identification, explosives, weapons, munitions)
- Control of networks for moving personnel and logistics
- Acts of sabotage (in urban centers)
- Clandestine medical facilities

Underground members are normally not active members of the community and their service is not a product of their normal life or position within the community. They operate by maintaining compartmentalization and having their auxiliary workers assume most of the risk. The functions of the underground largely enable the resistance movement to impact the urban areas.

The Auxiliary

The auxiliary refers to that portion of the population that is providing active support to the guerrilla force or the underground. Members of the auxiliary are part-time volunteers who have value due to their normal position in the community. The auxiliary should not be thought of as a separate organization, but rather as a different type of individual providing specific functions as a component within an urban underground network or guerrilla force's network. These functions can take the form of logistics, labor or intelligence collection.

Auxiliary members may or may not know any more than how to perform their specific function or service in support of the network or component of the organization. In many ways, auxiliary personnel assume the greatest risk and are the most expendable element within the insurgency. These functions are sometimes used to test a recruit's loyalty prior to his exposure to other parts of the organization. The functions of the auxiliary can be likened to embryonic fluid that forms a protective layer, keeping the underground and guerrilla force alive.

Specific functions include:
- Logistics procurement (all classes of supply)
- Logistics distribution (all classes of supply)
- Labor for special material fabrication (false identification, improvised explosives, weapons, munitions)
- Security and early warning for the underground facilities and guerrilla bases
- Intelligence collection
- Recruitment of new members
- Couriers and messengers as part of a communications network
- Distribution of propaganda material
- Managing safe houses
- Transportation for logistics and personnel

Resistance Force Commands

Resistance force headquarters are traditionally referred to in U.S. Special Forces doctrine as area commands. If an area command is divided into subordinate units, these are referred to as sector commands. These terms are important in that they refer to a region more than a unit's size. These commands control all forces of the resistance within their areas of responsibility, to include the guerrillas and the underground, and are subsequently responsible for all functions of the organization. This enables these groups to operate in a self-sufficient and decentralized manner. In addition to the regional commands, the whole of the insurgency may receive guidance from a single body of leadership. If this body exists within the resistance area, it is referred to as the shadow government. It can reside within the guerrilla force or the underground. If a body of leadership resides outside the country, it is referred to as the government-in-exile.

How the Environment Shapes the Structure of an Insurgency

The environment significantly affects the structure or operational design of an insurgency. In some cases, the structure can be mistaken or mislabeled as a strategy. Planners and advisers need to have an appreciation for this difference, otherwise the plan for support and advice rendered may prove inadequate or insufficient.

The following excerpt from the unclassified 1989 RAND study, *The Nicaraguan Resistance and U.S. Policy*, highlights this exact problem with that particular campaign effort. "The resistance's lack of authenticity as an indigenous insurgency and the Contras' extreme dependence on U.S. support were deprecated even by participants who otherwise generally favored active U.S. support to anti-communist insurgencies. The resistance has always been structured (inappropriately) as a force with short-term, purely military objectives. The U.S. effort to assist the Contras in Nicaragua was obviously handicapped by a lack of expertise on how to effectively organize and prosecute an insurgency. The United States erred particularly in structuring the Contras as a conventional raiding force that depended heavily on outside resupply."

Understanding why one pattern works well in a given environment, but may not work in a different one, is the key to avoiding template-like solutions to complex and unique problems. Fundamental to the successful development of a UW strategy is a comprehensive understanding of how insurgencies function. According to a 1995 RAND study, *The Urbanization of Insurgency*, most successful insurgencies have been able to generate and maintain a rural and urban capability. In the 1966 DA PAM 550-104 *Human Factors Considerations in Undergrounds and Insurgencies*, a comparison of 24 case studies placed the average ratio between rural guerrillas and urban cell members at 1 guerrilla to 9 underground/auxiliary members. In agrarian based cultures, this gave the insurgency control over the rural areas, including the numerous small villages contained within and influence in the major cities. While this degree of balance is a desirable structure for a developing insurgency, the fact remains that certain environments do not support this design.

The significant rise in urbanization in the last 60 years has promulgated a change in modern insurgent strategies, increasing the importance of the urban aspects. This places as much if not more emphasis on the development of the underground as compared to the traditional guerrilla bands. In these cases, the ratio of

rural guerrilla to urban cell member may become as high as 1 to 30 underground/
auxiliary members. This should not be interpreted as an advantage as much as a
challenge. Without a rural capability, the ability of an insurgency to execute large-
scale attacks is greatly limited, thereby limiting the overall ability to challenge the
host-nation security apparatus. While urban insurgencies can certainly coordinate
and conduct dramatic small-scale attacks, such as bombings and stand-off rocket

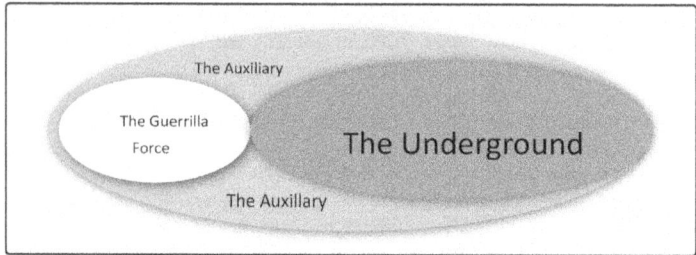

Figure 1. The traditional ratio of guerrillas to underground/auxiliary members

attacks, the psychological impact of guerrillas overrunning a host-nation police or
military outpost is a significant demonstration of power to a population.

In rare cases, where the operational environment is particularly remote and
devoid of a significant population, it may be possible for guerrillas to function almost
independently of an over arching resistance movement or insurgency. However, this
is more the exception than the rule, and this seemingly more familiar military-like
structure should not automatically be used as a template for other cases.

Figure 2. An example of an insurgency almost completely comprised of a guerrilla force

Phases of an Insurgency

There are various steps or phases that insurgencies traditionally progress
through during their development. Without an appreciation for these steps, military
planners can have unrealistic expectations for the time required to develop and employ a guerrilla offensive capability. The following six steps are intended to serve as

an example of how an insurgency may develop, with a specific focus on the preparations prior to the establishment of a guerrilla capability. These steps should not be confused with the phases of U.S. support to a resistance movement or insurgency, which will be addressed later in this document. The material below is largely derived from a 1961 document, *A Basic Doctrine for the Conduct of UW,* written by OSS and CIA veteran Frank Lindsay.[2]

The development of organized resistance in a territory must follow a carefully planned sequence of steps, each of which must be completed successfully before the following step can be safely taken.

1. Organization of clandestine networks. The first essential is to be able to move resistance organizers and propaganda material into the territory; support them clandestinely, and provide for continuing communications with adjoining resistance bases. These networks also pave the way for later active resistance by developing active support from a significant part of the total population.

2. Organization of intelligence and counterintelligence networks. No resistance organization can operate successfully unless it is fully protected by a deep intelligence screen. Equally, it is necessary to establish effective counterintelligence nets to verify every participant's loyalty to the movement and, through penetrations into the enemy security services, to identify those who are acting as double agents.

3. Organization of local area commands in areas of favorable terrain. Such area commands will be the forerunners of later fully-armed guerrilla units, and will provide the transition between the entirely clandestine civilian organization and the mobile armed guerrilla force living in forests and other inaccessible areas. These commands will be given full responsibility for the organization of their geographical areas, including the organization of supplies, intelligence, propaganda and communications. As the area commands continue to develop, they may further subdivide the area into subordinate sector commands. These sector commands will carry the brunt of attacks against enemy installations and forces, while drawing support from the area commands.

4. The development of direct communications with the outside. At approximately this point in the development of resistance in the area, direct communication with the outside world can be established, although it may be established in some cases at an earlier stage.

5. Gradual development of free territory in the hinterlands. As the fighting units further develop in strength, it will be possible to establish relatively large areas that are normally free of any enemy troops and all enemy controls over the civil population. At this point, these "liberated" territories can begin to prepare and organize to replace the enemy structure on the completion of the conflict.

The principal danger at this stage is that the resistance leadership will become overconfident and will become too quickly adjusted to the relative tranquility of life in territories from which the enemy has been expelled. If this occurs, the resistance movement will likely suffer serious setbacks as a result of violent enemy counterattacks for which it will find itself unprepared.

A second major danger at this stage is the over-recruitment of guerrilla forces. It will be a natural tendency on the part of a previously uncommitted civilian population, seeing the apparent success of the resistance forces, to attempt to join the resistance as its final success appears assured. There will also be a tendency on the part of guerrilla leadership to accept all recruits in order to deny them to the enemy, and to incorporate them into the organized structure of the resistance. This can cause serious losses as there is a maximum strength of resistance forces that any particular territory can absorb and provide with room for dispersal. Guerrilla forces in excess of this upper ceiling will contribute little to the fighting strength, and become a heavy burden on the command for supply and for protection in the event of strong enemy attack.

6. Final consolidation of formerly separated liberated areas. It is at this point that the former clandestine area commands transform into the overt governing authority.

U.S. Army Special Forces doctrine addresses phases of an insurgency as a construct for describing the current state of an insurgency in a very precise manner. These doctrinal phases are the latent incipient phase, the guerrilla warfare phase and the war of movement phase. While the previously mentioned six steps are somewhat more detailed than this doctrinal three-phase model, Steps 1-4 essentially represent the latent incipient phase, Step 5 represents the guerrilla warfare phase and Step 6 represents the war of movement phase.

The Development of a Resistance Movement Following an Invasion

During the aftermath of an invasion, pockets of potential resistance normally form out of groups that either deliberately went into hiding or have inadvertently become isolated from the main society by the invading forces. The structure and organization of these groups will likely be very ad-hoc and disconnected, as compared to an insurgency that has developed over a decade due to dissatisfaction with the status quo of the state. These forces have not had the benefit of time to develop their plan or seek the skills required to resist. In this situation the value of understanding the normal progression of an insurgency becomes more evident. When assessing a situation or developing a plan to support a given group, planners and advisers must be able to recognize when steps have not taken place or have taken place in a less than effective manner.

Personnel who migrate toward the resistance are often individuals who feel they have the requisite skills to resist. Many occupations have significant experiences that translate well into resistance. Active and retired military, police and hunters have the skills needed to function as guerrillas. Business connections, religious institutions and community organizations can form the basis for a network. Although not formally trained, criminal elements are very familiar with techniques such as operating through compartmentalized networks and countersurveillance. The challenge these groups will face is in adapting their existing skills to the new unfamiliar environment and against an organization not bound by the rules of the former state. They will have to adapt while applying these skills during an unforgiving period of trial and error.

These forces are likely to experience a sense of isolation that will impact their psyche and decision making. The enemy will attempt to portray their situation as hopeless and futile and possibly offer a viable alternative to rejoin the new society.

Personnel not prepared for the stress associated with life as a resistance member are very susceptible to "giving up." It is exactly at this point, while allied forces, in conjunction with the government in exile *slowly* develop a plan to reclaim the lost territory, that a message must be sent to the population now living in occupied territory; "Do not lose faith. You have not been forgotten. Your government still exists. Help is on the way!"

If these pockets can survive long enough to establish supporting clandestine infrastructure and coordinate their efforts with other dislocated pockets, they might be able to transform into an actual resistance movement. It is the enemy's challenge to consolidate control and eliminate these pockets before they mature. External sponsorship, particularly in the form of professional assistance (in the form of trainers and advisers) and critical logistics can serve as the catalyst that ignites a smoldering resistance potential into an actual capability.

How the U.S. Military Conducts UW

U.S. support to an insurgency can be categorized as one of two types of campaign efforts: as part of general war scenarios and limited war scenarios. When a UW campaign is conducted in support of an eventual conventional invasion as part of a general war scenario, the objectives and goals focus primarily on operational and tactical military objectives. When a campaign is conducted in support of a limited war scenario (not overtly involving U.S. forces), the objectives take on a more strategic political and psychological nature that normally outweighs the operational and tactical military objectives.

UW In Support of General War

The purpose of this type of operation is normally:
- To facilitate the eventual introduction of conventional allied forces.
- To divert enemy resources away from other areas of the battlefield.

The normal phases for U.S. support for this type of operation are below in Figure 3. The progression of these phases with the operational signature is depicted in the graph below (Figure 4).

Examples of this type of UW effort by the United States include:
- The OSS in the European theaters
- U.S./Filipino resistance (1942-45)
- North Korea (1951-53)
- Cold-War contingency plans for Europe (1952-1989)
- Kuwait (1990-91) *(aborted)*
- Afghanistan (2001-2002)
- Northern Iraq (2003)

During these types of operations, the resistance forces can assume a greater degree of risk by exposing nearly all its infrastructure based on the expectation of success and link-up with coalition invasion forces (Fig. 4). If the intent of the UW operations is to develop an area in order to facilitate the entry of an invasion force, the challenge is to ensure that the resistance operations complement rather than inadvertently interfere with or even compromise those of the invasion forces. If the timing is wrong or the conventional invasion forces fail to liberate the territory and link up with resistance forces, it is likely that the resistance organization (guerrillas, underground and auxiliary personnel) will suffer significant losses.

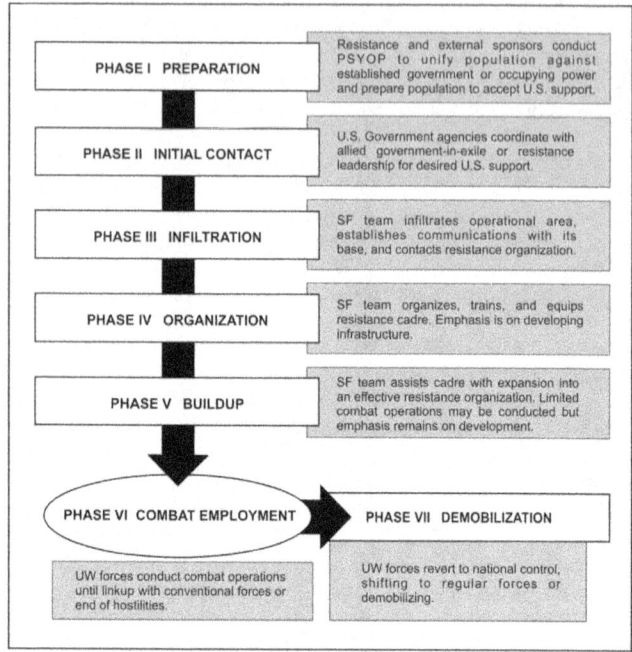

Figure 3. The seven phases of U.S. sponsorship for UW

In this scenario, UW planners should **not** know the specific date of D-Day (if it is even known at the time of planning for support to resistance forces). At some point these personnel may be infiltrating into enemy territory. This type of information is far too sensitive to be known by personnel in these positions due to their potential to become captured and subsequently jeopardize the entire invasion. UW planners need to know parameters that would allow them to plan accordingly without compromising the pending invasion details. An example of this could be *"no earlier than 120 days after C-day (C+120) a capability to disrupt the enemy in sector XXXXX must be in place and ready for activation. BPT to sustain combat operations for 30 days after the commencement of open hostilities by the U.S. ground forces."* This does not imply that D-Day is C-Day +120 but rather that a window for when D-Day could occur begins at C+120.

Additionally, an agreement must be reached between the UW headquarters, or HQ, and the conventional HQ regarding how the UW force will be notified of the commencement of D-day. This is critical in order for the U.S. advisers and/or the resistance force to be able to alert and mobilize their operational forces. These forces have planned numerous compartmentalized operations, ranging from guerrilla attacks, sabotage, reception of conventional forces and deception. Many of these elements have been standing by waiting for an initiation signal. They may require

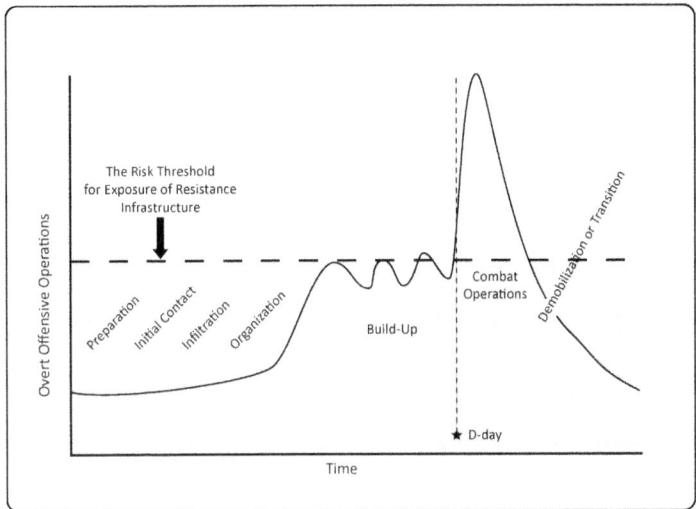

Figure 4. The phases of U.S. sponsorship in support of a conventional allied invasion

a period of darkness to alert, assemble force and materials and move into position prior to execution. An example of how this might be written in an order could be "*execute operations (the specific type TBD by you once on the ground) to disrupt the enemy at H-hour, D day to D+2. You will be notified 24 hrs in advance of the required effects.*" Needless to say, the need to balance operational security and still achieve synchronization is particularly challenging, especially if both sides are unaware of these challenges and have never practiced this previously.

Wide-scale resistance operations are not well suited for main battle areas. They are more effective along the flanks and rear of the enemy area. With this in mind, resistance forces can perform considerable supporting operations that range from intelligence collection, reception for conventional advance airborne and amphibious forces such as pathfinders and scout swimmers, deception operations, sabotage and limited interdiction.

If the intent of this type of UW effort is to compel an adversary to commit his forces to another part of the battlefield away from the proposed invasion area this is reasonably simple with a few exceptions. The challenge in this scenario is "Which resistance actions will trigger the desired responses? "How much is enough?" and "When to begin combat operations to appropriately affect the adversary's decision cycle?" If these operations are not coordinated with the invasion force or if they are timed incorrectly they can have significant unintended negative consequences. It is not difficult to imagine unintentionally causing a negative effect by damaging a bridge intended for later use by the allied invasion force, or alerting an enemy with a premature spike of resistance activity too far in advance of an invasion.

If guerrillas start to achieve a degree of parity with the host-nation forces, particularly if overt U.S. support in the form of Special Forces advisers becomes permissible, their tactics may change to allow them to fight as partisan units, rather than as guerrilla bands waging guerrilla warfare. U.S. advisers need to be careful that this transformation does not occur prematurely and inadvertently draw their respective guerrilla units into a decisive engagement that they are not prepared for largely due to the belief that close air support will make up for other deficiencies. Planners should not become fixated on only using the resistance force as security for the U.S. advisers in order to provide terminal guidance for air interdiction missions. While this is a viable application, when fully developed and employed, the capabilities of the resistance forces can have a much broader physical and psychological impact on the enemy.

It is the responsibility of U.S. Special Forces personnel to ensure a successful link-up between conventional forces and the resistance forces in order to prevent fratricide and facilitate a rapid advance of the invasion forces. Once this occurs, Special Forces personnel will be in a position to focus and coordinate the return to normalcy in the urban centers through the resistance force's temporary facilitation of basic civil infrastructure and governance roles.

Following successful link-up with the conventional forces, they can perform a host of manpower-intensive security tasks, particularly in the rear and flank areas away from the front lines. During the Normandy invasions, resistance forces sabotaged rail cars intended to deliver German armor units to the beachhead, and attacked hundreds of small targets ranging from attacks on logistics facilities, check points, lines of communication and leadership. Their networks actively recovered downed air crews and transported them to safety until they could be returned to advancing allied units. Following the invasion, they were immediately integrated into local security functions.

UW in Support of Limited War

The purpose for this type of operation is usually to apply pressure against an adversary of the United States. If U.S. adviser participation is permissible this type of campaign could also be conducted with the seven-phase model. However, if direct U.S. advisers' involvement is limited to a training and advisory role from across a friendly border, the traditional phases are adjusted. In place of Phase 3 (Infiltration), indigenous personnel are removed from the occupied territory, receive training and are reinserted back into the operational area (possibly to serve as cadre). *(The reader should be aware that this nine-phase model is not doctrine as opposed to the seven-phase model.)* The nine-phase model is:

- Preparation
- Initial contact with resistance force representatives
- Removal of selected indigenous personnel from the operational area
- Training for selected indigenous personnel
- Reintroduction of indigenous personnel into the operational area
- Organization of the resistance forces
- Build-up of resistance force capability
- Sustained combat operations
- Transition to normalcy

Without the introduction of conventional invasion forces, the nature of protracted combat operations is very different from the seven-phase model mentioned previously. The insurgency must consider the acceptable level of exposure of their organization in relation to the enemy's willingness and ability to retaliate. The insurgents must ensure they retain the initiative at all times and never expose too much of the organization. These types of operations are highly sensitive and normally conducted with a relatively small U.S. footprint, often in theaters that do not have a higher joint task force.

With this type of effort, it is likely that the level of involvement by U.S. advisers in denied territory will be limited due to the political risk involved. In these instances the U.S. advisers' level of influence over the insurgent group's action will be less than if they were participating alongside the resistance forces in combat operations (like in the seven-phase model). For this reason, it is particularly critical that the ideological goals of the U.S. and the group receiving support must be aligned. If this is not the case, the relationship will prove unproductive in the long term. It is critical not to overfocus on short-term tactical objectives at the expense of long-term impact on the region.

The normal progression of the phases with the operational signature is depicted in the graph below (Figure 5).

Examples of this type of UW effort by the United States include:
- OSS in the Asian/Pacific theaters (1943-1945)
- The Baltic States (Estonia, Lithuania, Latvia) (1950s)
- Guatemala (1954)
- Albania (1949-54)
- Tibet (1955-61)

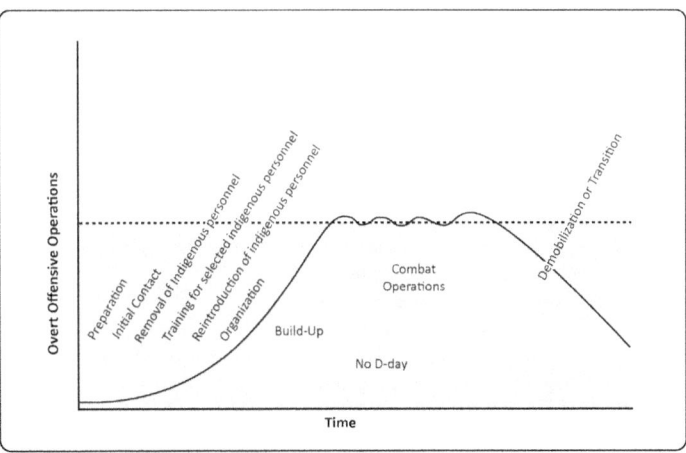

Figure 5. The phases of U.S. sponsorship when large-scale involvement is not anticipated

- Indonesia (1958)
- Cuba (1960s)
- N. Vietnam (1961-64) *(aborted)*
- Afghanistan (1980s)
- Nicaragua (1980s)

The Roles of the Army, Air Force, Navy and Interagency

UW is not restricted to any single military service; however, specific aspects have historically been the responsibility of each service.

The Army has traditionally provided the trainers and advisers to resistance forces. This specifically equates to training or advising guerrilla units in guerrilla warfare and underground organizations, including their respective auxiliary personnel, in the full spectrum of required functions, such as guerrilla warfare, sabotage, subversion, intelligence collection and logistics.

Psychological operations units assist and facilitate the tactical and operational HQ in the production and distribution of media intended to bolster the morale of the resistance and population or undermine the governing authority's legitimacy, whether a host-nation government or occupying military force. This support can take the form of audio broadcasts, Internet and printed material. Subversion is a critical line of operation in this type of campaign and should not be viewed as merely a supporting task assigned to the psychological units. The psychological operations personnel and units provide expertise for the HQ in how to achieve the desired results. This is largely done from outside the denied territory and coordinated through the advisers on the ground with the resistance forces.

As a collateral activity, U.S. personnel can establish capabilities with the resistance forces to assist in the recovery of friendly personnel, particularly downed aircrews in the course of evading, or post escape, in enemy-controlled territory. This activity is referred to as unconventional assisted recovery. During World War II, this capability rescued thousands of downed airmen. Although still a viable task with resistance forces, improvements in fighter and bomber aircraft technology, coupled with the introduction of the helicopter in the last 60 years have reduced the criticality of this capability to a fraction of its former historic requirement.

The Air Force and the Navy have traditionally provided infiltration, exfiltration and resupply support. In the absence of normal lines of communications and supply, the importance of this logistical support cannot be overstated.

Numerous other government agencies, such as the Department of State and the CIA, also play important roles in supporting UW efforts.

Prerequisite Conditions for Successful UW

The successful execution of UW is contingent on certain prerequisite conditions in the environment, some of which are beyond control and some that can be influenced. If these factors are not clearly understood by planners or overlooked by decision makers, the likelihood for missed opportunities, inappropriate application or unintended consequences will be high. It is essential that leaders and planners maintain objectivity in their analysis of these prerequisites. They must remain cautious of becoming overly focused on seeking opportunities to conduct a *desired* type of operation rather than determining the type of operation needed to achieve the desired end state. U.S. military doctrine has traditionally stated, "Special Forces do not create resistance movements, they support them." This is not a reflection of a lack of authority, but rather the reality that existing resistance potential can be developed further, but it cannot be manufactured.

In April of 1961, two weeks before the Bay of Pigs incident, Lindsay wrote a document entitled *A Basic Doctrine for the Conduct of UW*.[3] This document was based on his significant experience supporting the Yugoslav Partisans while with the OSS in World War II and various unsuccessful covert efforts with the CIA to promote resistance in Soviet satellites during the early days of the Cold War. One of the most critical aspects of his document was the realization that conducting UW in peacetime against an adversarial regime requires different operational techniques than those utilized during general war against occupying armies. From his experience, he identified the following prerequisites for the successful prosecution of UW (the following section on prerequisites comprises the original work of Lindsay with the exception of my bold comments).

A Weakened or Unconsolidated Regime

The organizational mechanisms by which the regime in power maintains control over the civilian population must be sufficiently unconsolidated or weakened to permit the successful organization of a minimum core of clandestine activities. It will be almost impossible to organize successful resistance under a fully consolidated government. Unless the regime has some chinks in its control system, any clandestine activity will become penetrated and controlled by the enemy before it attains the minimum size and dispersal necessary for survival and effective action.

A Population of Strong Character

The character of the population must be sufficiently strong to sustain resistance over long periods through serious reprisals, and against the repressions of the regime in power. Populations that have recently been overtaken by an occupying military force have a very different character than those that have had to survive for decades under an oppressive regime. Over time, populations can lose their ethnic, cultural and religious identities to those imposed by an oppressive regime. This was the case in some former Soviet satellites and potentially the case in regions under a religious extremist ideology.

Favorable Terrain

The terrain must provide a minimum of cover as protection against surprise attack from the enemy. Normally this will mean forest cover or mountains and relatively large areas free from intersecting roads and rail lines. Resistance can take place in cities, but it has a special character and will usually require support from major resistance centers located in the hinterlands where terrain is more favorable. Resistance activities in cities will normally be limited to clandestine activities and to unarmed, overt civil actions such as mass demonstrations and strikes. Resistance in cities can, in turn, provide important support to armed guerrilla activities in the hinterland.

Since the original writing of this document certain changes have occurred that impact this prerequisite and warrant mentioning. Advancement in technology, such as UASs and the helicopter, have reduced the potential safe havens in restrictive terrain traditionally utilized by guerrilla elements. Since the end

of World War II, guerrilla elements have increasingly had to seek safe havens across sympathetic international borders or in artificial favorable terrain in the form of ghettos and refugee camps. Unlike actual favorable terrain, such as jungle-covered mountains, these artificial safe havens only continue to provide utility as long as the host-nation security forces are willing to tolerate them.

Compatible Goals and Ideology with Those of the United States

The goals of a resistance movement must essentially be indigenous goals. It is unlikely that these goals will be identical with those of the United States. Nevertheless, through the skillful influence of the U.S. representatives with the resistance leadership, a greater community of interest can be established, which will make it possible for the United States to provide assistance and to establish substantial influence over the course of the resistance itself. It is essential to recognize that the only goals and objectives that will provide sufficient motivation for a successful resistance are those that develop indigenously, or are soundly based on indigenous political and social conditions, and that goals and objectives artificially imposed from the outside will not find sufficient acceptance to make for a strong and successful resistance.

Through timely preparation the United States can influence the selection and emergence of the indigenous leadership and can, within limits, influence the objectives, organization and conduct of the resistance movement. The time to exercise this influence will be before the resistance begins and during its early stages. By the time a resistance movement gains sufficient strength to be "visible" to the world, it will be too late to exercise very much influence over the character and goals of the movement and over the selection of leadership.

It is essential to recognize that leadership rarely can be imposed from the outside, either in the form of U.S. representatives or in the form of expatriates who are sent back by the United States. The leadership must arise internally, and must prove itself at every stage in order to justify internal confidence and support it will need to be successful.

Capable and Skilled Indigenous Leadership

The resistance leader must first define his political goals in terms that can win the support of most of the population. He must communicate these effectively to the majority of the population in an environment in which all of the overt forms of communication are controlled by his enemy. He must therefore be able to establish a pervasive clandestine underground organization so that he can reach, with his propaganda, a majority of the population; so that he is able to be informed at all times of the movements of enemy forces and of the plans being made against him; so that he can supply his forces clandestinely by civilian contributions of food, clothing and money; and so that he can recruit into the active fighting arm of his movement the necessary manpower to ultimately prevail. In doing this, he will have no organized system of taxation and supply. He will have no established courts, police forces and prisons with which to enforce his orders. He will have no organized system of propaganda and no organized recruiting service to supply him with manpower.

The resistance leader must be able to combine divergent political groups and objectives into a single common program to defeat the enemy. From the standpoint of U.S. interests, he must not only be able to do all of these things, but he must have the political astuteness to lay solid foundations during the resistance period for the type of postwar government that will be consistent with long-range United States interests. Resistance leadership must, from the outset, plan and prepare for the ultimate assumption of power. Many of the decisions that will determine the success or failure of the resistance leadership in establishing an effective postwar government will be made irrevocably during the resistance period. Failure to plan and prepare for the postwar period may thus jeopardize the realization of postwar aims. Decisions postponed until the war is won may be too late.

Because internal resistance breeds such violent emotions and antagonisms, and because it encourages and depends on defiance of established order, a long period of resistance and guerrilla warfare will make the postwar re-establishment of an organized society infinitely more difficult. It must, therefore, be a duty of resistance leadership to build during wartime toward the establishment of a postwar society in which various elements and groups have not become so violently antagonistic to each other that they cannot work together in the postwar period. Similarly the wartime leadership must, with the help of its external allies, train during the war a full cadre of technical and administrative personnel so that they will be able to step into positions of civil administration and labor in order to rapidly restore essential social services and infrastructure.

In most cases, successful wartime resistance leadership will become the postwar government. The time to influence the character of the postwar government is, therefore, at the earliest stage in the organization of resistance.

The Criticality of the Feasibility Assessment

Planning must remain limited until certain assumptions have been confirmed as valid. If operations proceed without a proper assessment of feasibility, the likelihood of unintended consequences is high. To gain an accurate picture, operational personnel will need to meet with indigenous personnel who represent the resistance forces. This can take place inside the denied territory (extremely high risk), in the U.S. or in a third-party nation. While meeting representatives in the U.S. or a third nation provides a safer option for an assessment team, it conversely provides a less reliable assessment for potential capabilities.

The normal questions that comprise a feasibility assessment are as follows:

• Are there groups who could be developed into a viable force?

• Are we in contact with or can we make contact with individuals representing the resistance potential in an area?

• Are there capable leaders, with goals compatible with U.S. goals, which are willing to cooperate with the U.S.?

• Can the leaders be influenced to remain compliant with U.S. goals?

• Are their tactics and battlefield conduct acceptable by the Law of Land Warfare and acceptable to the U.S. population?

• Will the environment (geography and demographics) support resistance operations?

• Does the enemy have effective control over the population?

• Is the potential gain worth the potential risk? Is this group's participation politically acceptable to other regional allies?

Expatriates can prove to be a valuable resource, particularly in regions where the culture is largely unfamiliar or alien to a planner's frame of reference. However, great care should be taken to ensure the individual's claims are valid. An expatriate's influence in a given country is inversely proportional to the length of time he has been away from his former homeland and the level of control measures, propaganda and intimidation employed against the population. While there are many reasons an expatriate might exaggerate his influence in a region and attempt to exploit the situation in his favor, he may legitimately be surprised to find his own assessment of his influence to be grossly inaccurate. During normal peacetime conditions, a person can spend years away from a country and expect to maintain his contacts and influence. Under the pressures of a harsh regime or occupation, this time period is reduced significantly.

Operational personnel involved in determining the feasibility of a potential campaign must have (1) clear campaign objectives, (2) a desired end state and (3) knowledge of exactly what level of support is available and acceptable. Without these specifics, negotiations with potential resistance forces would be futile. During assessment, if conditions prove to be unfavorable, planners need to also consider if there are measures that could change the current situation to one that would be favorable. For example:

- Can a potential resistance group be persuaded to cease unacceptable tactics or behavior?
- Can a coalition ally be persuaded to accept a specific resistance group's participation under certain conditions?
- Can the enemy's control over the population be degraded?
- Can the population's will to resist be bolstered?
- What can actually be achieved given the constraint of time?

Operational detachments need time to organize with their indigenous counterparts, to develop a working relationship in terms of trust and credibility, and to build up the guerrilla capability and supporting infrastructure, while remaining relatively undetected by the enemy. These objectives would take considerable time to achieve in friendly territory, operating with U.S. units. For forces working within enemy territory, dealing with unfamiliar units and coordinating operations across a wide, decentralized and compartmentalized front, the time requirement is much greater.

Risk Analysis and Risk Acceptance

Planners and commanders need to appreciate the relationship between risk and capability. The resistance capability that can be developed is directly proportional to the amount of time available to operational detachments on the ground. If the risk associated with inserting U.S. operational detachments is considered to be unacceptable until the night prior to an invasion, the desired operational capabilities will likely not be in place for several months.

During operational phases, forces (to include U.S. advisers) are normally out of range of many capabilities that are generally accepted as the norm in conventional military operations. These absent capabilities may include medical evacuation, close air support and continuous lines of communication. The associated risks from not having these inherent standing capabilities can be mitigated to some degree by many SF operating techniques. Commanders will likely need to accept a greater degree of decentralization than they may be used to from operational elements during

these types of operations.
Logistics Challenges
Logistics support for UW has some unique challenges as compared to conventional operations. Guerrilla forces do not operate on a contiguous battlefield or with standard lines of communication. They must sustain themselves largely with material obtained from their environment, whether gained from the population or by battlefield recovery. External logistical support from a sponsor should only serve as a supplement to existing resistance logistics. Over dependence on external logistics could hamper the insurgent leadership's desire to develop and employ traditional underground and auxiliary supporting mechanisms. While this may seem simpler in the short-term, remaining disconnected from the population also causes the population to become disconnected from the insurgency. If conducted in a manner that doesn't overburden the supporting population, allowing local populations to provide support for the insurgency, although operationally challenging, serves to maintain a vital connection between the two entities.

While external support can provide a significant increase in capability, it comes with the risk of exposing the resistance force during its delivery and dispersion, not to mention the risk to the U.S. platform. Advisers need to carefully consider factors such as which items of supply warrant utilization of limited U.S. resupply platforms and the optimum time for their delivery. The delivery of logistics is one of the most powerful tools available to the adviser in his quest to develop influence. Early in a relationship, the adviser's future status can rest on the successful or unsuccessful delivery of supplies. It is entirely possible that a planned resupply may not take place due to the tactical situation, the weather or recent losses of platforms. Sometimes resupply efforts are unsuccessful because bundle parachutes fail to open, materials are dropped in the wrong location or bundles can contain the wrong materials. While none of these factors are the fault of the adviser, he alone has to deal with the consequences.

If an external sponsor provides logistical support, it must be compatible with the existing indigenous equipment. Requirements may dictate that the support is delivered in a manner that does not expose the operation. This can be a significant challenge for U.S. planners in that the requirements may be for non-standard materials and equipment that do not exist within the U.S. military supply system. Organizations with a UW mission requirement should be familiar with the unique procedures for the procurement, requisition, delivery and accountability of standard and non-standard types of material.

Command and Control Challenges
Communications

Conventional commanders need to be aware that tactics and techniques for command and control used in other operations will not apply during UW missions. During the conduct of UW, units will not be able to communicate with their headquarters in the same manner as during other types of operations. Even if the communication architecture is available, great care should be exercised before emplacing requirements on units operating from within enemy territory. Unlike conventional units, UW organizations risk some degree of exposure with every communication. Communications encryption should not be confused with not emitting a communications' signature. The U.S. military's insatiable appetite for computer-based briefings and real-time communications needs to be managed in relation to the operational environment. Units engaged in UW should always operate under the assumption that the enemy is listening to all open communications as well as conducting direction finding for unusual signals in urban and rural areas.

Managing the Size and Location of the U.S. Headquarters

Unlike conventional operations, the acceptable size and optimum location for headquarters of units engaged in UW change as the mission progresses. Whether a headquarters provides C2 from an adjacent country or chooses to infiltrate the resistance area, should be based on where it can provide the most value to the operation. SF company and battalion headquarters are tactical elements intended to not only coordinate the actions of their subordinate units, but also integrate with their resistance force counterparts (*sector or area command*) if needed. Unit headquarters need to consider (prior to entering into the guerrilla-controlled territory) if their signature will jeopardize the mission and if there is an appropriate level of the resistance's headquarters that could benefit from their direct interaction. In either case, large unwieldy headquarters would prove inappropriate for these types of operations, particularly in forward areas.

If an SF company, battalion or group chooses to infiltrate a small forward headquarters into the resistance area, they essentially relinquish or greatly degrade their ability to provide C2 for other special operations being conducted outside the area of operations. If an SF company headquarters is waiting to infiltrate into the occupied territory (or guerrilla-controlled territory), they normally do not simultaneously provide C2 for their detachments already operating in the resistance area. Operational control is retained by the SF higher headquarters and transferred to the company pending successful infiltration and link-up with resistance forces. At some point, the full battalion or group HQ might be able to operate in the resistance area, but this would likely be in the final phases of a UW campaign, when overt participation has become permissible.

In instances where large-scale U.S. involvement is not envisioned as part of the campaign, UW efforts are, more frequently than not, compartmentalized efforts executed by relatively small organizations, consisting primarily of senior personnel. Because of the political implications of such operations, the headquarters of the ground advisory forces needs routine direct access to the top U.S. decision makers for the campaign. This is particularly important if the U.S. advisers are directly interacting (and often negotiating) with senior resistance force leaders that regard the government-in-exile as their higher chain of command. It would be highly ineffective to institute an ad-hoc intermediate military chain of command of senior ranking officers in order to replicate a more familiar C2 structure.

Complex Relationships with Foreign Allies

UW efforts are generally launched from across the borders of a neighboring country to the targeted area. These countries may act as overt allies (as was the case of Great Britain in World War II) or act as neutral nations and conceal their cooperation (as was the case with India and Tibet in the 1950s, Honduras and Nicaragua or Pakistan and Afghanistan in the 1980s). These complex allied relationships are a reality in this type of operation and must be accepted and dealt with rather than avoided. In this regard, the State Department and Embassy Country Teams play a vital role in synchronizing efforts in these adjoining countries and must be well integrated with the executing UW HQ.

The resistance forces should be thought of as a partner ally. To view the resistance forces during a UW effort as anything less than an allied partner is both unrealistic in terms of the amount of control that can be exerted over them or an indication that the force is more a manufactured group of mercenaries rather than a true resistance force. In most cases it is unlikely that an insurgency or resistance is one homogenous organization. It is more likely a loosely aligned confederation of organizations that have probably never met before and have never seen the whole of their own organizations.

The U.S. Army Special Forces Group is specifically designed to synchronize the efforts between displaced sector and area commands with their Operational Detachment-Alphas (ODAs), Operational Detachment-Bravos (ODBs or company HQ) and Operational Detachment-Charlies (ODCs or battalion HQ). Each one of these Army unit HQ maintains the ability to operate with its equivalent resistance force counterpart.

Preparatory Measures Taken Within an Allied Nation at Risk

In areas that have a high potential of invasion, the U.S., alongside the allied partner, could take preemptive measures to prepare an area for future resistance efforts. In these cases, the U.S. could train host-nation military forces in resistance tactics and techniques with the intent that they would become the "pockets of resistance" that the U.S. would contact and support in the event of an invasion.

In these scenarios, planners need to avoid considering pre-existing pockets of dissident activity or resistance potential unless it is believed they could be persuaded to join into a coalition with the government in exile. Although these elements might provide some degree of resistance potential, their existing adversarial relationship with the allied government would negate this tactical value. Only resistance forces believed to be loyal to the existing government (potential government in exile) should be considered in scenarios with allied nations at risk of invasion. Consideration and contact with potential resistance forces without the allied partner-nation's knowledge risk inadvertently undermining the ally's legitimacy and damaging, if not ending, the partnership with the U.S. government. In the event of any U.S.-sponsored resistance training with host-nation military counterparts, it is critical that U.S. trainers and planners maintain the highest degree of confidence in the training recipients that they would not employ these skills against their own government at some later time.

In extreme cases, the U.S. could develop the basic supporting human and logistical infrastructure in a region with the potential to become a future pocket of resistance. Plans could be developed for stay-behind elements (U.S. and/or host nation) to be pre-positioned in a region anticipated to be overtaken by enemy forces. These groups would later serve as the cadre or nucleus from which resistance forces would be organized. These activities would be conducted in secret from the population, but with assistance of selected host-nation forces.

If these activities were conducted with the host nation, they would likely be perceived as a sign of U.S. commitment to that ally's survival and the development of a potential means to provide assistance, even if they lost control of their state to an invader. Conversely, if preparations for potential future resistance were conducted without the knowledge of the host nation and discovered, it would likely be perceived as a lack of confidence in that state's government and their continued relationship with the U.S. The value of unilaterally developing supporting infrastructure (*for resistance operations*) for potential future contingencies would need to be seriously considered in comparison to the actual incurred risk to the existing diplomatic relationship and legitimacy of the host-nation government.

Preparatory Measures Focused Toward a Potential Adversary

Preparatory actions focused toward a potential adversary are somewhat different than those activities conducted with an allied nation at risk of invasion. While many of the same activities are still applicable, in these cases it could be permissible to identify possible elements within a region with the potential to effectively develop into a resistance organization. This is significantly different than any UW preparations that might be taken within an allied country at risk of invasion (i.e. alongside the allied government forces). In this scenario, whether or not these groups would be willing to engage in a potential relationship with the U.S. government is less relevant than developing a basic understanding of these organizations to include their leadership, ideology, objectives and capabilities. It may also be permissible to determine host-nation military and security capabilities and weaknesses with regards to their control over the population.

These types of activities should only be considered in regions where the governments are categorized as "potential adversaries" rather than allies or neutral states. These actions would be counterproductive or even divisive to other U.S. government efforts at maintaining or strengthening an existing tenuous relationship. This course should only be considered if the relationship is deemed to be unrecoverable.

The Impact of Technology on UW

The significant increases in military technology impact how U.S. forces could support an insurgency or resistance movement. Certain capabilities, such as air support, if made available to guerrilla forces, can in some cases allow them to cease fighting as guerrillas and openly challenge the opposing military forces. While it might seem like an attractive option to rapidly raise the ability of the guerrilla force to openly challenge the opposing forces, if not managed correctly, it can have several unintended effects.

The first concern is the premature or inadvertent exposure of U.S. sponsorship. There are several examples where advanced military capabilities were injected into a supposedly indigenous insurgency, in lieu of developing the insurgency's internal capability, only to prematurely expose U.S. sponsorship. Any technology introduced into this type of campaign needs to be compatible with the rest of the existing strategy. It should augment and support the strategy and not make the strategy in support of the capability.

The second concern is the unintended transformation of the guerrilla force into make-shift infantry battalions. Advisers need to guard against overconfidence among guerrilla commanders who feel that, based on their previous success fighting as guerrillas, their units are capable of fighting as regular infantry...and that the U.S.-provided air support will make up for any deficiencies. Some planners may see this as an attractive option to increase the number of "battalions" in the fight. It's important to maintain realistic expectations for guerrilla forces. If guerrilla units place too much faith in U.S. air support to "carry the day" or are asked to operate in regions that they have no familiarity with or connection to the local population, the results will be poor.

It is not the goal of U.S. advisers to transform their counterparts into infantry battalions, but rather to assist their commanders in employing them in the most effective manner possible. The lure of technology can cause planners to mistakenly believe that transforming all these forces into infantry-like units is a feasible option. While it is difficult to quantify the psychological impact of wide-scale resistance, of which guerrilla attacks are only a part, the overall value cannot be overstated, but it can certainly be undervalued. Planners should possess a familiarity with the concepts and techniques of insurgency in order to synchronize this disruptive potential alongside conventional operations rather than default to transforming it into something that appears more familiar, but is in fact much less useful than its full potential.

Law of Land Warfare Challenges

The Law of Land Warfare outlines criteria for resistance forces and foreign advisers. In order for resistance forces to receive legal belligerent status as prisoners of war they must comply with the following four criteria:

- Carry their arms openly
- Wear a distinctive insignia
- Follow a chain of command
- Conform with standards of conduct outlined by the Law of Land Warfare

When irregular forces do not comply with these criteria, they do not qualify for legal belligerent status and are subject to the normal criminal laws of the government they are operating against. With regards to resistance movements, the Law of Land Warfare only addresses guerrillas as possible legal belligerents. This presents a problem for other members of the resistance, specifically underground or auxiliary members. These individuals are traditionally charged with espionage and often sentenced to long prison sentences or even death for their crimes against the state.

In order for foreign advisers to qualify for legal belligerent status, they must comply with the previously mentioned four criteria with the exception that they must wear the uniform of their own country. Depending on the mission and adversary, there are situations where qualifying for such status may not be a high priority. Under certain conditions, units conducting UW may deliberately choose to assume risk and forfeit their right to these qualifications for operational reasons. If and when this decision is made, it is a command decision and not an individual preference.

Some of these criteria are open to a degree of interpretation, particularly in terms of what exactly constitutes a distinctive unit insignia and what constitutes carrying arms openly. Regardless of "interpretations," if U.S. advisers learn of resistance behavior that would constitute a violation, such as human-rights abuses, they have a responsibility to attempt to correct these actions and report the behavior to their higher headquarters. The level of influence exercised by U.S. advisers will be proportional to the amount of direct U.S. support and their involvement in resistance-force operations. If U.S. advisers are accompanying guerrillas on operations, they are in a much stronger position to influence the guerrillas' behavior as compared to if they are restricted to bases in safe havens.

Advisers need to always remain cognizant of their own safety. If the resistance force perceives that the advisers are reporting negatively against them and possibly jeopardizing the continuation of support, a small group of advisers could suddenly find themselves in significant danger. Advisers must always consider the benefits versus the risks of maintaining and continuing a relationship with groups whose behavior is questionable. While disengagement might seem ethically and morally appropriate, remaining engaged affords advisers the potential to favorably alter a group's behavior for an operational and strategic long-term outcome.

U.S. UW Efforts from 1951- 2003
U.S. Military and CIA in Korea (1951-1953)

The U.S. military was largely caught off guard by the conflict in Korea. Despite tremendous success during World War II, the capability to support resistance movements was almost completely removed from within the Department of Defense by 1950. A fraction of the capability was retained in the newly established CIA, but proved to be inadequate to meet the needs of a full-scale campaign.

Overall this campaign effort was effective, but not nearly as effective as it could have been. The inability to establish sustainable guerrilla areas on the mainland hampered long-term operations. The U.S. provided personnel who were not familiar with the concepts and techniques associated with resistance operations. The military's best attempt to find a matching experience set was to send basic-training instructors. Operations made wide use of displaced refugees as recruits and were generally launched from outside enemy-controlled areas and resembled temporary raiding parties.

The resulting operations caused the military to re-think its position on a military force trained for this type of operation. In 1952, the first official U.S. Special Forces organization was established. In 1953, some of the newly developed Special Forces deployed to Korea. However, they were not employed as intended and generally used as replacements.

CIA in Albania and Latvia (1951-1955)

The CIA used these opportunities to test theories for rolling back communist domination of Eastern Europe. These locations were chosen due to their relatively small size. Both efforts were failures. From these efforts the CIA theoretically learned three lessons. First, techniques used during wartime do not apply in peacetime (emphasizing the need for covert methods of operating). Secondly, tyrannical regimes that have had the benefit of years to consolidate power have a much greater hold over the population than a newly occupying power. Subsequently, there must be a weakness that can be exploited. This specifically equates to the ability of the state to exert control over the population and the population's willingness to resist. Lastly, prior to any guerrilla operations, a sufficient base of support must be present in the form of auxiliary and underground networks, although the least visible and understood, it is the most time-consuming and difficult to establish. Developing guerrilla elements is relatively easy; keeping them alive is much harder. Trying to jump ahead to developing guerrillas is like trying to build the upper floors of a house before laying the foundation.

CIA in Guatemala (1954)

In 1954, various influential businesses persuaded members of the media as well as members of the U.S. government that the Arbenz government, which was the elected government of Guatemala, was leaning toward communism. The U.S. government, through the State Department and the CIA, convinced several senior Guatemalan military officials to overthrow Arbenz's government. This effort was supported by a highly successful psychological campaign that portrayed the overthrow as part of a much larger movement. Although widely suspected, the U.S. government's hand in this effort remained ambiguous for a couple of years until it was finally exposed as the follow-on governments were accused of being inept dictators by the Guatemalan population. Although often portrayed as a success, the hindsight of history now reveals the Guatemalan operations were in fact *"shameful, particularly so because the five decades of governments that followed the 1954 coup were far more oppressive than Arbenz's elected government. Aside from the morality, there were other unfortunate legacies of the Guatemalan "success". Allen Dulles used it as a model in advising President Kennedy seven years later to pursue the ill-fated Bay of Pigs invasion of Cuba."*[4] The Guatemalan effort serves as an example of tactical success in isolation of any wider sound strategy, particularly in the form of long-term effects on a region. This effort coupled with the coup of the elected Mossedegh government in Iran (1953) provided a strong argument to de-legitimize any U.S. claim to support democracy for the remainder of the Cold War.

CIA in Tibet (1955-1969)

While this campaign was very well executed, no level of tactical success in Tibet could defeat the Chinese. The National Volunteer Defense Army (NVDA) did successfully delay the Chinese victory by several years. In 1960, the first U2 aircraft was shot down. Although not related to the covert support efforts, the subsequent compromise of U.S. clandestine overflights of communist territory caused the U.S. President to suspend all further clandestine overflights, which essentially crippled the air resupply to the Tibetan guerrillas. A year later, the catastrophic failure of UW efforts in Cuba nearly ended the Tibetan program. The last resupply airdrop to Tibetan resistance forces took place in 1965. Although the U.S. ceased all support to the Tibetan resistance in 1969 as a condition for eventually establishing diplomatic relations with China, the Tibetan resistance continued to resist into the 1970s.

Two lessons can be drawn from the Tibetan efforts. As an alternative to prior U.S. efforts where advisers infiltrated into the combat area, in this effort selected personnel were extracted, trained and reinserted with reasonable success. The nature of the offensive operations in this type of campaign is different compared to operations that culminate in support of a TBD D-day. Without the eventual introduction of conventional forces (i.e., a D-Day) the combat operations need to be sustained for an undetermined period of time and therefore conducted in a manner that does not compromise the supporting resistance infrastructure.

CIA in Indonesia (1957-1958)

Fearing that the Indonesian government was leaning toward communism, after a declaration of neutrality, the CIA supported several Indonesian military officers who claimed to control a rebel army made up of former military units. During this operation, the CIA chose to employ refitted B-26 bombers and P-51 fighter aircraft flown by mercenaries from various countries. The rebel army had little popular support among the population while the Indonesian government still held control of the majority of its military. During the course of the support effort, direct American support was exposed several times causing embarrassment to the United States.

This situation finally came to a head when a B-26 flown by an American CIA contract employee was shot down and the pilot was captured by Indonesian forces. Although the CIA was confident that no link could be made between the U.S. government and the pilot, he in fact had documents on him that indeed did link him to Clark Air Force Base. The incident had been a major scandal for the Eisenhower administration. Soon after the scandal broke, the CIA ceased all support and the Indonesian government crushed the rebellion with conventional military forces. A particularly interesting point in this case is that Indonesia was not considered a belligerent state when the decision was made to support the rebels. The U.S. State Department maintained relations with the government and had an embassy in Jakarta during the operation.

CIA in Cuba and the Bay of Pigs (1961)

The CIA applied a variety of the nine-phase model, consummate with covert operations, but attempted a culminating D-day style uprising rather than sustained guerrilla warfare. There were some reasonably successful efforts to develop underground elements, but these efforts were not coordinated with the larger campaign. They did not employ appropriate techniques to conceal the operation or the U.S. participation. The "resistance movement" was generally manufactured, rather than fostered from an already existing one in country. Lastly operational security in the United States was a complete failure.

The lessons of the Bay of Pigs are as follows:

• Resistance forces require clandestine infrastructure and support mechanisms.

• These operations require planners and advisers who understand the dynamics of insurgency/ resistance operations (including guerrilla warfare and underground operations and how they integrate with each other).

• When there is a population involved, the psychological piece (i.e. propaganda and subversion) may be more important than the physical piece (guerrilla warfare).

• This type of effort requires the ability to conduct the full spectrum of non-attributable psychological operations in conjunction with the physical piece.

• Any organization involved in UW (particularly peacetime) needs a comprehensive understanding and process for oversight, C2 and OPSEC for covert and clandestine ops.

CIA and Special Forces in Laos (1959-1962)

In 1959 the United States began a secret program called White Star. This operation was coordinated in conjunction with the French government to train Laotian Army Battalions. Personnel came from the 1st and 7th Special Forces groups. In December of 1960, a group of Laotian paratroopers attempted a coup d'etat that caused the withdrawal of the French advisers, whom had previously restricted American military involvement to the role of trainers. During the counter coup, Special Forces personnel advised the remaining parachute forces to rout the rebel paratroopers, winning great favor with the Laotian government. By 1961 the original contingent of nine ODAs had grown to 21 ODAs and the effort expanded to include an advisory role. The counterinsurgency efforts also included training for selected tribal elements in civil defense. A concept was developed to expand the tribal operations [covertly] to the rural areas in northern Laos parallel to the North Vietnamese border. *Special Forces also undertook two major unconventional programs in northern Laos (as compared to the other counterinsurgency programs). Both programs concentrated on training minority tribal groups as irregular forces capable of conducting guerrilla warfare across rugged terrain or behind the enemy lines. The Kha program, the second unconventional training activity, was designed by Special Forces to use the various Kha tribes in harassing and raiding enemy rear bases and installations especially along the Ho Chi Minh trail, the primary overland North Vietnamese resupply route across Laos into South Vietnam and Cambodia. Unfortunately higher authorities refused to adopt the Special Forces suggestions regarding the newly created guerrilla movements and due to political decisions the Meo and Kha programs never realized their full potential.*[5] After 1962, the effort was dramatically reduced when Laos declared its neutrality. While the Special Forces involvement ended, U.S. support continued covertly by the CIA to General Vang Pao's "Army Clandestine" for another decade.

The counterinsurgency programs in Laos are sometimes misrepresented as UW. The counterinsurgency efforts conducted against the Pathet Lao in southern Laos were significantly different from the proposed UW efforts envisioned for northern Laos and North Vietnam. The proposed UW efforts were intended to interdict and harass regular North Vietnamese forces who were occupying Laotian territory in order to establish supply networks into South Vietnam which later became the Ho Chi Minh trail.

CIA and Special Forces in North Vietnam (1961-1964)

In 1960, the CIA began attempts to establish agent networks within North Vietnam. The program, called Leaping Lena, was highly penetrated by double agents and never produced any viable results beyond a handful of questionable agents. The hopes of conducting UW against North Vietnam had diminished by 1964 when the military assumed control of the operation. The Special Observations Group's or SOG's actual name had been UW Task Force. The Special Forces concluded Leaping Lena by parachuting the questionable trainees over North Vietnam and terminating further support. From this point on operational efforts focused on portraying a false resistance movement intended to confuse the North Vietnamese.

The program shifted to covert coastal raids (similar to North Korea) and covert reconnaissance and interdiction missions into Laos and Cambodia. **Special Forces in South Vietnam (1957-1975)**

As a result of the increased concern over communist subversion in Vietnam and at President John F. Kennedy's insistence, SF added COIN to its list of missions in 1961, making a significant point that it is different from UW and not a subcomponent of it. The majority of Special Forces efforts in South Vietnam were not UW but rather COIN, specifically the Civilian Irregular Defense Group, or CIDG, the Provisional Reconnaissance Units, or PRU, and Mike Forces.

The CIDG were indigenous irregulars who provided a vital counterinsurgency role of area denial by securing hamlets.

The PRU were indigenous platoons developed as part of the CIA Phoenix program. These elements identified insurgent support infrastructure (or underground and auxiliary members). This effort could be called counterterrorism, or counterinsurgency.

SOG also developed platoon- and company-size strike forces called Mobile Guerrilla Forces and Bright Light teams. The Mobile Guerrilla Forces acted as quick-reaction forces for recon teams in contact with Viet Cong elements and the Bright Light teams conducted raids to rescue allied prisoners and downed air crews. The Mobile Guerrilla Forces conducted four static line jumps during their operations and the Bright Light teams conducted nearly 100 raids (in S. Vietnam, Laos and Cambodia). While these efforts utilized recruited indigenous personnel, they were not considered UW. In 1970, this program disbanded and transformed to an adviser program for Vietnamese and Cambodian counterguerrilla forces.

See Colonel (R) Al Paddock's (the author of *U.S. Army Special Warfare and Its Origins*) comments regarding a review of *Imperial Grunts* (and the misuse of the term UW) in the March-April 2006 *Special Warfare*. Paddock, a Special Forces veteran of SOG and the CIDG mission stated, *"In truth, the mobile guerrilla forces can be more likened to World War II long range penetration units such as Merrill's Marauders or Wingate's Chindits. This is not to say that the Mobile Guerrilla forces did not perform useful or heroic missions. They did, but not as guerrillas."*

CIA and Special Forces in Nicaragua and Honduras (1980-1988)

The United States supported various resistance groups that opposed the socialist Nicaraguan Sandinista government. These groups, which operated from Honduras and Costa Rica, collectively became known as the Contras. RAND after action reviews from this covert operation criticize it for being artificially manufactured and not having legitimate support inside Nicaragua. "This effort was developed almost entirely along military lines and advisors lacked practical understanding of the requirements to develop and run an insurgency." (*U.S. Support to the Nicaraguan Resistance*, 1989 and *The U.S. Army's Role in Counterinsurgency and Insurgency*, 1990). The Contras have been compared to paid mercenaries with no real political component and connection to the population. This may partially explain the lack of guerrilla bases inside Nicaragua. The other part of the explanation comes from the strength of the government counterguerrilla forces. The Contras would not have survived without the safe havens across neutral borders. Ultimately the covert operation, which became widely exposed due to the Iran-Contra scandal, ended badly for the U.S. government and caused feelings reminiscent to the post Bay of Pigs-era regarding covert support to insurgents.

CIA and Special Forces in Pakistan and Afghanistan (1980-1991)

While this covert effort remained primarily training and material support, it grew to the point that the potential risk of providing indigenous guerrillas with U.S.-manufactured lethal aid (such as Stinger anti-aircraft missiles) became acceptable. U.S. personnel established supporting bases in Pakistan (following the nine-phase model mentioned previously). There are several lessons of the operations in Afghanistan. Covert operations of this nature require neighboring partner countries to develop support bases. This necessity causes a level of compromise in U.S. objectives. Pakistan controlled the distribution of U.S. support to rebel groups and subsequently was able to alter the balance of power among resistance group in Pakistan's favor.

Afghanistan highlights the importance of maintaining perspective of long-term goals and not allowing enthusiasm to overtake decision making. The importance of a feasibility assessment before the decision to throw support at a group can't be overemphasized. Any support will change the balance of power and dynamics of the region. When a group is supported, engagement should continue until post-conflict stability is achieved. The Taliban came to power because the U.S. ceased support to Northern Alliance troops and other Islamic nations did not cease their support to these radical groups.

The sensitivity of these types of covert operations demands that operational commands interact directly with the national command authority and not through various "go betweens" in a long military chain of command.

Cold War Contingency Plans for Scandinavia and Europe. (1952-1989)

Although not executed, it is worth including the contingency plans that Special Forces would have implemented in the event of escalated hostilities in Europe during the Cold War. There is little debate as to the nature of what SF was prepared to do in this event; unfortunately, these plans count for little as they fade farther away with each passing day. There would be much value to declassifying some of this information for the betterment of the military's knowledge. At the height of the Cold War, 10th Special Forces Group remained prepared to employ 50 plus detachments across the whole of Europe. In addition to 10th SF Group, Det-A, although manned with SF Soldiers not officially part of 10th SF Group, remained prepared to conduct UW with its six detachments specifically in the urban areas of Berlin and northeast Germany.

Kuwait (1990-1991)

Following the Iraq invasion of Kuwait in August 1990, several isolated pockets of Kuwaiti resistance formed. Although the U.S. was in contact with elements of the Kuwaiti government in-exile in Saudi Arabia, efforts to support and coordinate the resistance forces was poorly integrated with the main military campaign planning. Planning efforts were too little and way too late. Although Kuwait offered little favorable terrain to support resistance operations, ad hoc resistance forces did operate without U.S. support until their eventual destruction. The operation may have proven unfeasible, but other circumstances rendered that debate irrelevant. Post-war after-action reviews indicated a general lack of organic capability, a lack of understanding of the requirements for supporting a resistance throughout the Department of Defense (to include special-operations forces) and a lack of synchronization between DoD and the interagency.

Afghanistan (2001-2002)

Between October 19 and November 20, TF Dagger (5th Special Forces Group) inserted 11 ODAs that linked-up with and coordinated the actions of Tajik, Uzbek, Hazar and Pashtun bands, loosely referred to as the Northern Alliance, in their efforts to defeat the Taliban government, which was providing safe haven to Osama bin Laden. These efforts were overwhelmingly successful. It is worth noting that the general weakness of the Taliban as a military force and the dramatic increase in lethal capability by the integration of U.S. air power created a state of parity between the Northern Alliance and Taliban not typical of UW efforts. Success was largely due to the operational capabilities of the inserted personnel. In many cases, these personnel were familiar with the culture, the region and, in some cases, possessed regional language skills.

Iraq (2002-2003)

As part of the invasion of Iraq in 2003, two UW campaigns were considered: A northern effort with the Kurdish Resistance predominately made up from the Patriotic Union of Kurdistan (PUK) and the Kurdistan Democratic Party (KDP) and a southern effort comprised of various Shia resistance groups. For a variety of reasons, the southern effort was not executed. The example of Iraq is significant because it demonstrates a scenario where two UW efforts could have been conducted in support of the conventional campaign, but each for entirely different purposes. The northern effort was intended to keep the Iraqi forces focused on the northern part of the country, making them unavailable to counterattack the coalition's southern invasion. The southern effort would have been executed to facilitate the introduction of invasion forces. While many of the environmental conditions were not optimal for coordinated resistance efforts in the south, there was also a perception among military planners that the potential value that might be gained by enabling Shia resistance forces would be unnecessary for the success of the coalition invasion and therefore did not warrant the possible associated risk.

The northern effort was considered the third priority for special operations, until Turkey refused U.S. forces access to bases, eliminating a second conventional maneuver force from the north. Without additional U.S. conventional forces invading from Turkey, only six U.S. divisions would participate in the initial invasion from the south. Roughly 30,000 Kurdish forces, supported by two battalions of Special Forces, successfully tied down 12 of Iraq's 20 divisions during the invasion and liberated the cities of Kirkuk and Mosul.

Notes:

1 JP 3-05 Doctrine for Joint Special Operations, 17 Dec 2003.

2 Lindsay, Franklin, *A Basic Doctrine for the Conduct of UW*, April 1961, accessed through the John F. Kennedy Library archives.

3 Ibid.

4 Accessed via https://www.cia.gov/library.

5 *Special Forces at War in SE Asia,* 1957-1975, Shelby Stanton, p. 22.

**U.S. ARMY JOHN F. KENNEDY
SPECIAL WARFARE CENTER AND SCHOOL**

**USAJFKSWCS
Bldg. D-3206 Ardennes Street
Ft. Bragg, N.C. 28310-5000**

http://www.soc.mil/swcs/swmag/

www.ingramcontent.com/pod-product-compliance
Lightning Source LLC
Chambersburg PA
CBHW050246230526
45470CB00005B/2130